I Know Shapes

Reading consultant: Susan Nations, M.Ed.,
author, literacy coach, and consultant in literacy education

Photographs by Gregg Andersen

Developed for Harcourt, Inc., by Gareth Stevens, Inc. This edition published by Harcourt, Inc., by agreement with Gareth Stevens, Inc. No part of this publication may be reproduced or transmitted in any form or by any means, electronic or mechanical, including photocopy, recording, or any information storage and retrieval system, without permission in writing from the copyright holder.

Requests for permission to make copies of any part of the work should be addressed to Permissions Department, Gareth Stevens, Inc., 330 West Olive Street, Suite 100, Milwaukee, Wisconsin 53212. Fax: 414-332-3567.

HARCOURT and the Harcourt Logo are trademarks of Harcourt, Inc., registered in the United States of America and/or other jurisdictions.

Printed in China

ISBN 13: 978-0-15-360214-6
ISBN 10: 0-15-360214-7

11 12 13 0940 16 15 14 13
4500409957

Harcourt
SCHOOL PUBLISHERS

circle

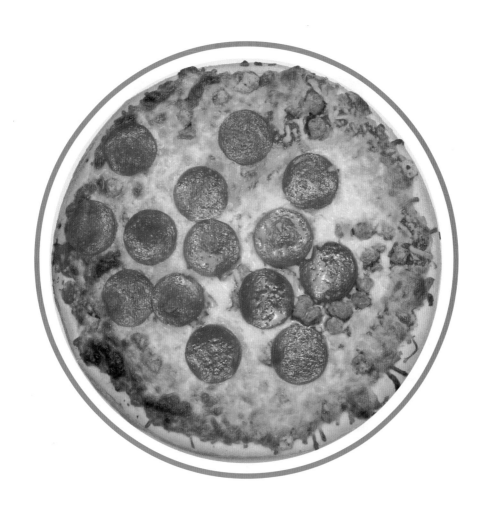

A circle is round.

side

corner

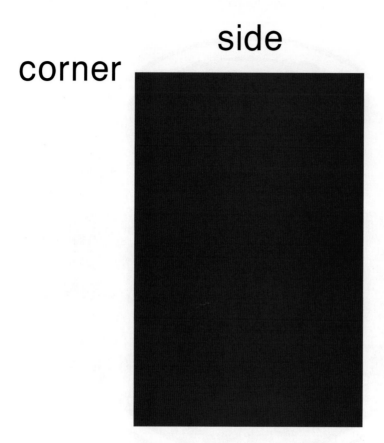

rectangle

A rectangle has 4 corners.
It has 4 sides too.

corner

side

square

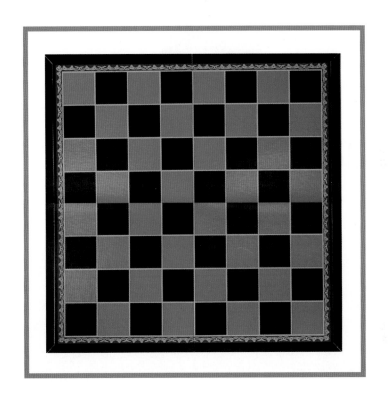

A square has 4 corners.
Its 4 sides are the same.

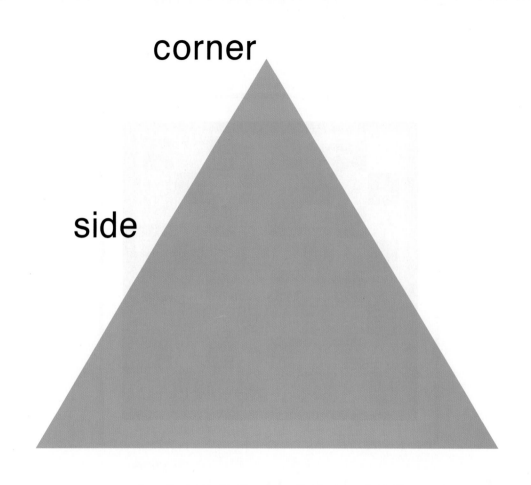

corner

side

triangle

8

A triangle has 3 corners.
It has 3 sides too.

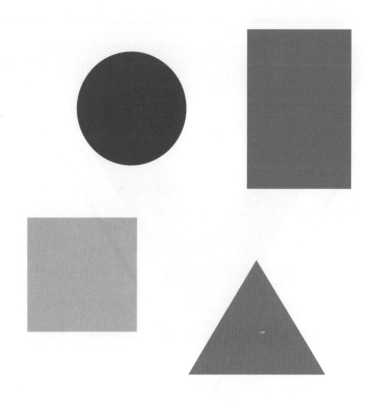

Shapes are fun!

Glossary

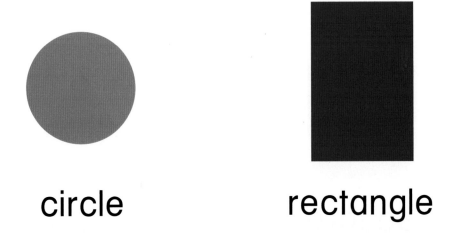

circle   rectangle

Glossary

square

triangle

Think and Respond

What shapes do you see?

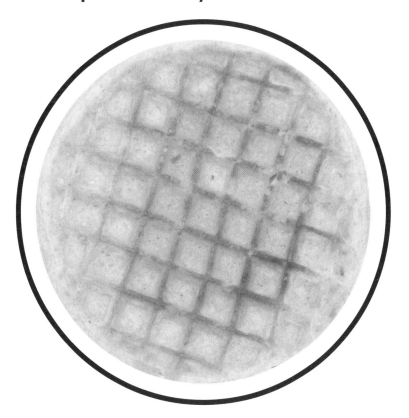

GRADE K

WORD COUNT

48

GENRE

Expository Nonfiction

LEVEL

See TG or go Online

Harcourt Leveled
Readers Online Database
www.eharcourtschool.com

ISBN 13: 978-0-15-360214-6
ISBN 10: 0-15-360214-7

90000 >

9 780153 602146

Developed for
Harcourt School Publishers by
GARETH**STEVENS**
GS
CLASSROOM

Harcourt
SCHOOL PUBLISHERS

I Know
Alike and Different

Harcourt
SCHOOL PUBLISHERS